SUPER-DEADLY ANIMALS

BLACK RABBIT BOOKS

JOANNE MATTERN

TABLE OF CONTENTS

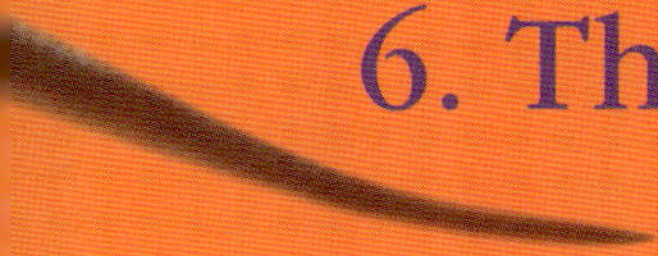

1

The Gila Monster

Be careful in the desert! Gila (HEE-lah) monsters live in the southwestern United States and Mexico. Most lizards can't hurt you. But the Gila monster has a secret weapon.

These lizards have glands in their bottom jaw. The glands are full of venom. The Gila monster bites. Venom flows into its victim.

The bite hurts a lot. Its venom usually can't kill a person. But it does kill the small animals it eats.

Did You Know? Poisonous animals are dangerous when touched or eaten. Venomous animals bite or inject toxins.

2 The Poison Dart Frog

These pretty frogs come in lots of colors. Don't be fooled. Those bright colors are a warning. The frog's skin is covered with poison. Just touching one can kill you.

Poison dart frogs live in the rainforest. Bugs there eat toxic plants. The frog eats the bugs. Then it stores those toxins in its body to make poison.

The golden poison frog is the most deadly. It has enough poison to kill 10 men.

Think About It

Poison dart frogs born in zoos are not poisonous. Why do you think this is true?

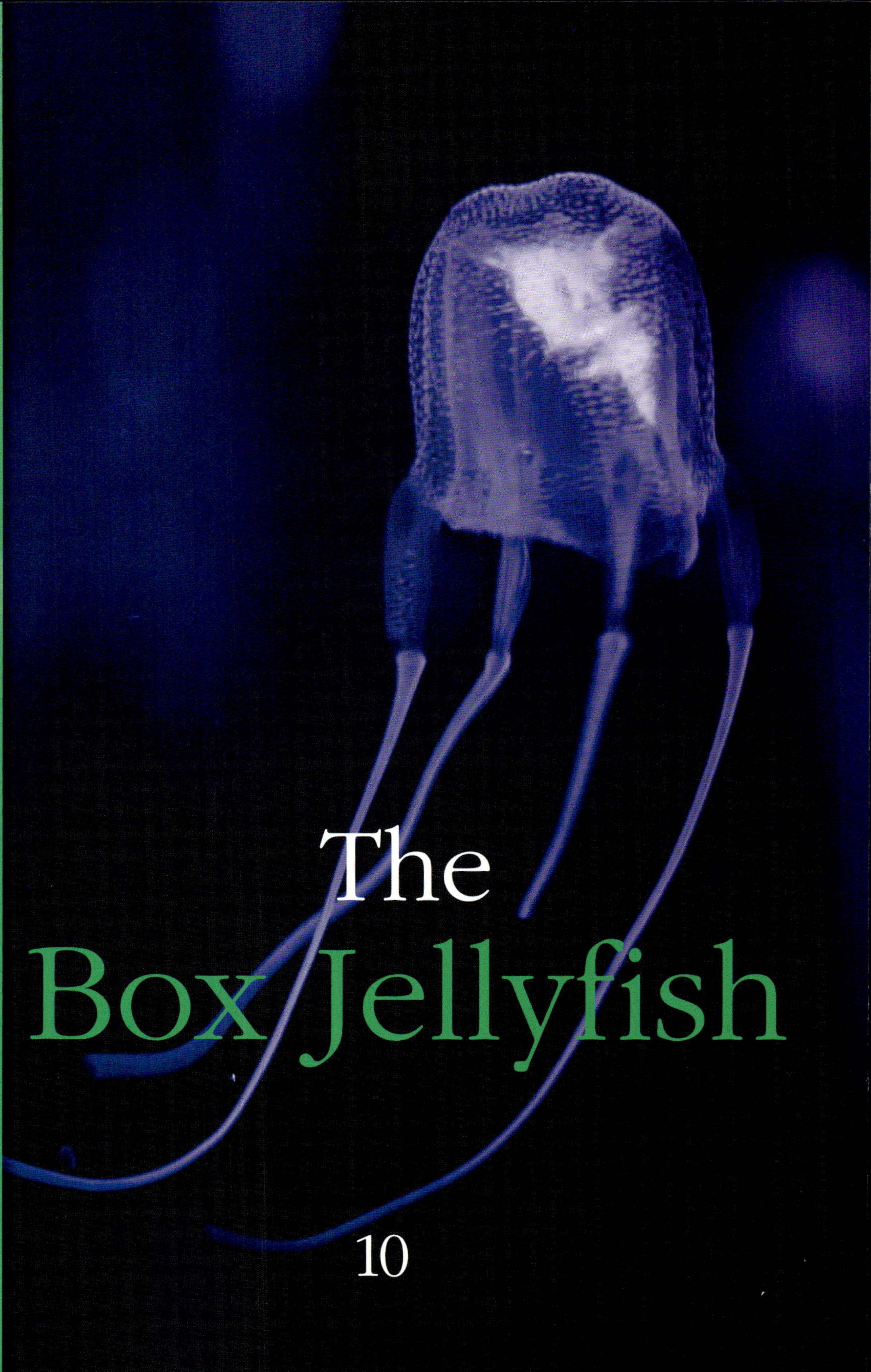

3

The Box Jellyfish

Box jellyfish do not look dangerous. But their long tentacles are a deadly weapon. They are covered with up to 5,000 cells. Each cell is full of venom. That venom is one of the deadliest in the world. Just one sting can kill a person in minutes.

Box jellyfish live in warm oceans. They have 24 eyes. They hunt shrimp and other small animals.

Did You Know?

Most jellyfish just float.
But a box jellyfish
can swim.

4

The Inland Taipan

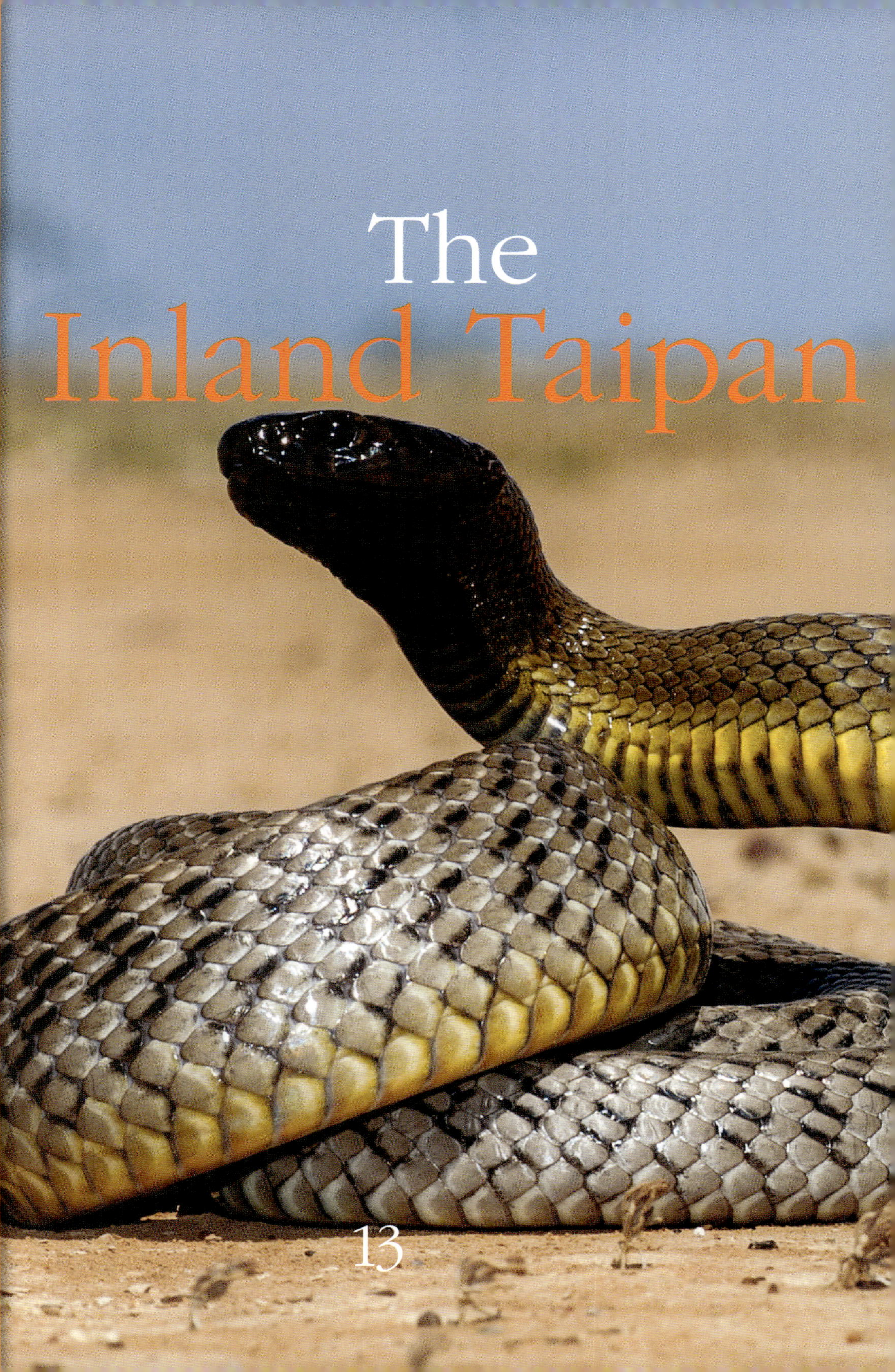

Many snakes have venom. But the inland taipan is the most deadly. This snake lives in the Australian outback. One bite has enough venom to kill 100 men.

Here's some good news. Inland taipans are very shy. They are not likely to attack people.

A taipan's prey is not so lucky. Rats are its favorite food. The snake corners a rat. Then it bites. Its venom works quickly. The rat cannot fight back.

Think About It

Think about where the inland taipan lives. What could be another reason this snake rarely bites people?

5

The Funnel Web Spider

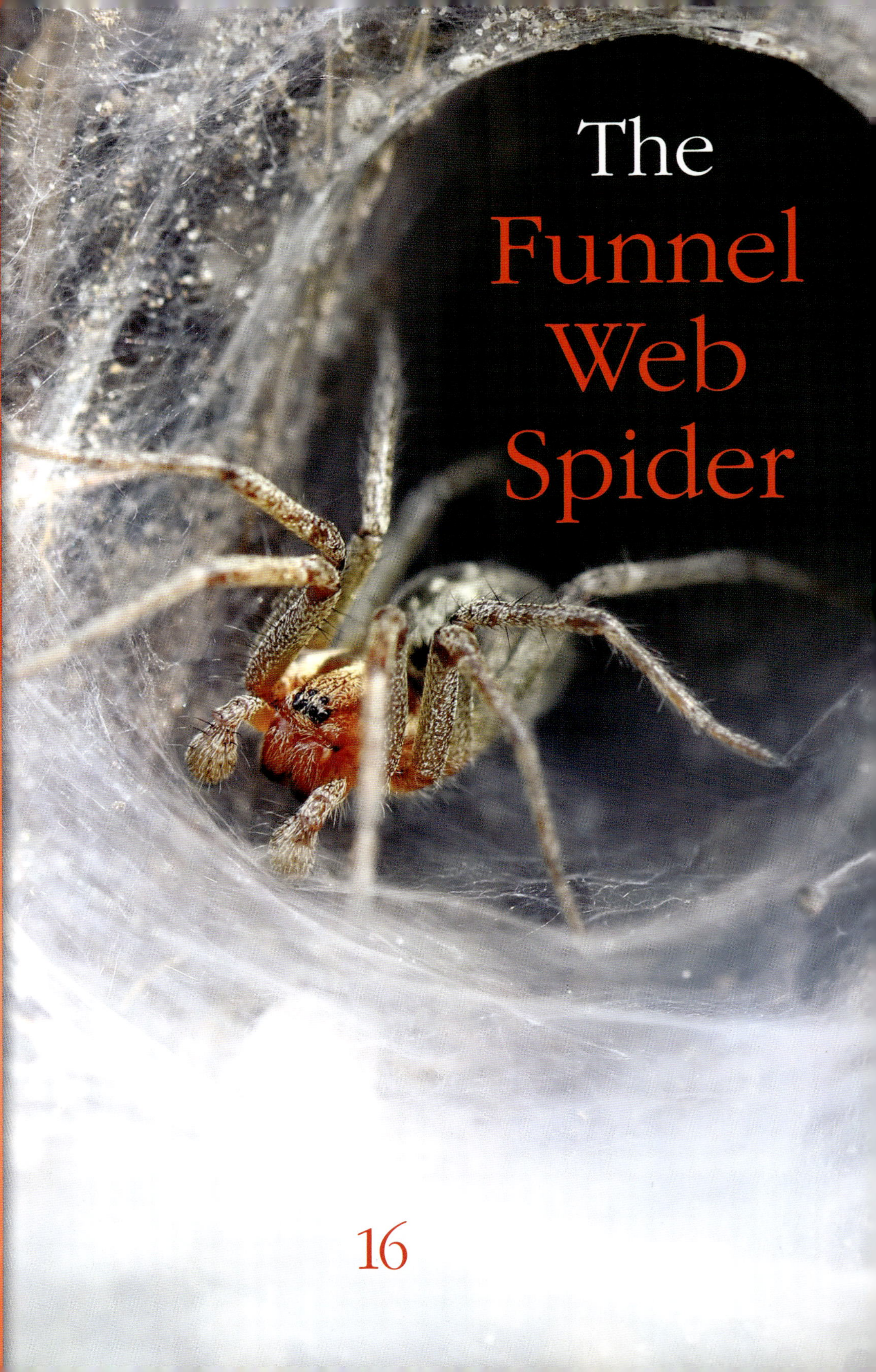

Australia is full of deadly animals. The funnel web spider is another one. This spider has strong venom. It can kill a person in 15 minutes.

This spider eats bugs and frogs. The venom **paralyzes** its prey. But something else happens when it bites a human. The person's heart pounds. They can't breathe. Soon, they are dead.

Funnel web spiders like damp places. They usually live under rocks. Some live in trees.

Think About It

Some animal venom is used to make medicine. What are some other ways deadly animals can help people?

6

The Blue-Ringed Octopus

This sea animal might look cute. But it is one of the deadliest animals on Earth. The blue-ringed octopus makes a toxin called TTX. The octopus bites. Venom flows. It can kill a person in 20 minutes.

The octopus's body is filled with TTX. If you eat it, you could die. So, this animal is both venomous and poisonous.

This creature eats crabs and shrimp. Its mouth is shaped like a beak. This beak breaks through the crab's tough shell. Then the octopus injects its venom.

Did You Know?

The blue-ringed octopus is the size of a golf ball.

MORE TO EXPLORE

WHERE IN THE WORLD?

GILA MONSTERS
North America

POISON DART FROGS
Central and South America

BOX JELLYFISH
Indian and Pacific Oceans

INLAND TAIPANS
Central Australia

FUNNEL WEB SPIDERS
Coast of Australia

BLUE-RINGED OCTOPUSES
Indian and Pacific Oceans

MORE TO EXPLORE

FANTASTIC FACTS

A Gila monster often chews its prey. Chewing injects more venom into the bite.

Poison dart frogs got their name from the Choco people. They put its poison on the darts they used for hunting.

The largest box jellyfish lives in the ocean around Australia.

Inland taipans are also called fierce snakes.

A funnel web spider's fangs can bite through fingernails.

A blue-ringed octopus changes color when it is scared.

MORE TO EXPLORE

COOL COMPARISONS

How deadly are the animals in this book? Let's rank them.

1 Inland Taipan
Enough venom to kill 100 men.

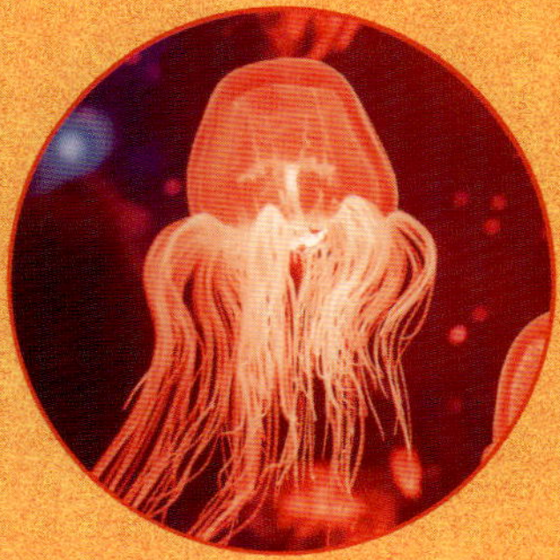

2 Box Jellyfish
Enough venom to kill 60 men.

3 Blue-Ringed Octopus
Enough venom to kill 26 men.

4 Poison Dart Frog
Enough venom to kill 10 men.

5 Funnel Web Spider
Can kill a person in 15 minutes.

6 Gila Monster
Not usually fatal to people, but very painful.

MORE TO EXPLORE RESOURCES

Glossary

gland (GLAND) A group of cells or an organ that produces fluids that are released into the body or pass out of the body.

outback (OWT-bak) Remote and rural areas, especially in Australia.

paralyze (PAR-uh-lahyz) To make powerless or unable to act, function, or move.

prey (PRAY) An animal hunted or killed for food.

tentacle (TEN-tuh-kuhl) A long thin body part on the head or around the mouth of some animals. Tentacles are used for feeling or taking hold of things.

toxin (TOK-sin) Any poisonous substance that is produced by living cells or organisms.

venom (VEH-nuhm) A poison made by animals used to kill or injure.

Read More

Beer, Julie. *Bite, Sting, Kill: The Incredible Science of Toxins, Venom, Fangs, and Stingers.* Washington, DC: National Geographic Kids, 2023.

Musgrave, Ruth. *Poisonous and Venomous Animals.* New York: DK Publishing, 2023.

Index

TOP RANK is published by Black Rabbit Books, P.O. Box 227, Mankato, MN, 56002.

• Top Rank is an imprint of Black Rabbit Books. • Edited by Alissa Thielges • Designed by Danny Nanos • Photographs © Dreamstime/Geoffrey Green, cover, Nh77, cover; iStock/Howard Chew, 20, 23; Shutterstock/Artur Janichev, 9, 21, Dewald Kirsten, 10, 21, GoodFocused, 2, 15, 23, James van den Broek, 18, JayPierstorff, 4, 21, Ken Griffiths, 13, 17, 21, 23, Kuttelvaserova Stuchelova, 14, lokesh164, 11, maniacpixel, 20, Marek Mierzejewski, 16, 21, Nuttawut Uttamaharad, 12, 23, Rosa Jay, 8, Tim Nolan, 6–7, 23, SChantra, 19, 21, Vladislav T. Jirousek, 5, 23, Zulashai, 11 • Printed in China

Library of Congress Cataloging-in-Publication Data: Names: Mattern, Joanne, 1963- author. | Title: Super-deadly animals / by Joanne Mattern. | Description: Mankato, MN : Top Rank is an imprint of Black Rabbit Books, [2025] | Series: Super-incredible animals | Includes bibliographical references and index. | Audience: Ages 8–11 | Audience: Grades 2–3 | Identifiers: LCCN 2024017516 | ISBN 9781645823810 (library binding) | ISBN 9781645824039 (paperback) | ISBN 9781645824251 (ebook) | Subjects: LCSH: Poisonous animals—Juvenile literature. | Venom—Juvenile literature. | Classification: LCC QL100 .M3755 2025 | DDC 591.6/5—dc23/eng/20240515 | LC record available at https://lccn.loc.gov/2024017516